# Start

# MATH PUZZLE-01

|   |   |   |   |   |   |   |   |     |
|---|---|---|---|---|---|---|---|-----|
| 2 | + |   | + |   | + |   |   | 26  |
| + |   | + |   | + |   | + |   |     |
|   | + |   | + | 16| + |   |   | 35  |
| + |   | + |   | + |   | + |   |     |
|   | + |   | + |   | + |   |   | 28  |
| + |   | + |   | + |   | + |   |     |
|   | + |   | + |   | + |   |   | 47  |
| 28|   | 37|   | 38|   | 33|   |     |

**Fill in the missing numbers**

The missing values are the whole numbers between 1 and 16.
Each number is only used once.
Each row is a math equation.
Each column is a math equation.
Remember that multiplication and division are performed before addition and subtraction.

# Exciting Math Puzzle For Adults
## Easy to Hard Addition, Subtraction, Multiplication & Division

**This book belongs to:**

_____

_____

### Thank you for purchase
Copyright and other intellectual property laws protect these materials. Reproduction or retransmission of the materials in whole or part thereof in any manner, without the prior written consent of the copyright holder, is a violation of the copyright law and will be enforced to the fullest extent of the applicable law.

# Mathematical Puzzles

**Fill in the missing numbers**

The missing values are the whole numbers between 1 and 16.
Each number is only used once.
Each row is a math equation.
Each column is a math equation.
Remember that multiplication and division are performed before addition and subtraction.

## Example

### Puzzle

|   | − |   | + |   | − | 6 | 10 |
|---|---|---|---|---|---|---|---|
| × |   |   |   | × |   |   |   |
| 1 | − | 11 | − |   | × |   | −30 |
| − |   |   |   | − |   |   |   |
| 8 | + | 7 | − |   | × |   | −193 |
| − |   |   |   | + |   | × |   |
|   | − | 14 | − |   | − |   | −13 |
| −14 |   | −1 |   | 46 |   | −16 |   |

### Solution

|   | − | 3 | + |   | − | 6 | 10 |
|---|---|---|---|---|---|---|---|
| × |   |   |   | × |   |   |   |
| 1 | − | 11 | − | 5 | × |   | −30 |
| − |   |   |   | − |   |   |   |
| 8 | + | 7 | − | 16 | × |   | −193 |
| − |   |   |   | + |   | × |   |
|   | − | 14 | − |   | − |   | −13 |
| −14 |   | −1 |   | 46 |   | −16 |   |

# MATH PUZZLE-02

## Fill in the missing numbers

The missing values are the whole numbers between 1 and 16.
Each number is only used once.
Each row is a math equation.
Each column is a math equation.
Remember that multiplication and division are performed before addition and subtraction.

# MATH PUZZLE-03

|   | + |   | + |   | + |   | 21 |
|---|---|---|---|---|---|---|---|
| + |   | + |   | + |   | + |   |
|   | + | 6 | + |   | + |   | 27 |
| + |   | + |   | + |   | + |   |
|   | + |   | + | 15 | + |   | 41 |
| + |   | + |   | + |   | + |   |
|   | + |   | + |   | + |   | 47 |
| 22 |  | 30 |  | 48 |  | 36 |   |

**Fill in the missing numbers**

The missing values are the whole numbers between 1 and 16.
Each number is only used once.
Each row is a math equation.
Each column is a math equation.
Remember that multiplication and division are performed before addition and subtraction.

# MATH PUZZLE-04

|   | + |   | + |   | + |   | 28 |
|---|---|---|---|---|---|---|----|
| + | ■ | + | ■ | + | ■ | + |    |
|   | + |   | + |   | + |   | 46 |
| + | ■ | + | ■ | + | ■ | + |    |
| 3 | + |   | + |   | + |   | 16 |
| + | ■ | + | ■ | + | ■ | + |    |
|   | + |   | + |   | + | 13 | 46 |
| 32 |  | 42 |  | 29 |  | 33 |    |

**Fill in the missing numbers**

The missing values are the whole numbers between 1 and 16.
Each number is only used once.
Each row is a math equation.
Each column is a math equation.
Remember that multiplication and division are performed before addition and subtraction.

# MATH PUZZLE-05

|   | + |   | + |   | + |   | 31 |
|---|---|---|---|---|---|---|----|
| + | ■ | + | ■ | + | ■ | + |    |
|   | + |   | + |   | + |   | 22 |
| + | ■ | + | ■ | + | ■ | + |    |
|   | + | 14| + |   | + | 16| 53 |
| + | ■ | + | ■ | + | ■ | + |    |
|   | + |   | + |   | + |   | 30 |
| 23|   | 46|   | 39|   | 28|    |

**Fill in the missing numbers**

The missing values are the whole numbers between 1 and 16.
Each number is only used once.
Each row is a math equation.
Each column is a math equation.
Remember that multiplication and division are performed before addition and subtraction.

# MATH PUZZLE-06

|   | + |   | + | 14 | + |   | 33 |
|---|---|---|---|----|---|---|----|
| + |   | + |   | +  |   | + |    |
|   | + |   | + |    | + |   | 38 |
| + |   | + |   | +  |   | + |    |
|   | + |   | + |    | + |   | 28 |
| + |   | + |   | +  |   | + |    |
|   | + | 3 | + |    | + |   | 37 |
| 43 |  | 35 |  | 35 |  | 23 |   |

**Fill in the missing numbers**

The missing values are the whole numbers between 1 and 16.
Each number is only used once.
Each row is a math equation.
Each column is a math equation.
Remember that multiplication and division are performed before addition and subtraction.

# MATH PUZZLE-07

|   | + |   | + | 12 | + |   | 31 |
|---|---|---|---|----|---|---|----|
| + | ■ | + | ■ | +  | ■ | + |    |
|   | + |   | + |    | + |   | 27 |
| + | ■ | + | ■ | +  | ■ | + |    |
|   | + |   | + |    | + |   | 41 |
| + | ■ | + | ■ | +  | ■ | + |    |
|   | + |   | + | 7  | + |   | 37 |
| 23|   | 29|   | 49 |   | 35|    |

**Fill in the missing numbers**

The missing values are the whole numbers between 1 and 16.
Each number is only used once.
Each row is a math equation.
Each column is a math equation.
Remember that multiplication and division are performed before addition and subtraction.

# MATH PUZZLE-08

|   | + |   | + |   | + |   | 31 |
|---|---|---|---|---|---|---|----|
| + | ■ | + | ■ | + | ■ | + |    |
|   | + |   | + | 3 | + |   | 44 |
| + | ■ | + | ■ | + | ■ | + |    |
|   | + | 8 | + |   | + |   | 30 |
| + | ■ | + | ■ | + | ■ | + |    |
|   | + |   | + |   | + |   | 31 |
| 39 |  | 45 |  | 25 |  | 27 |   |

**Fill in the missing numbers**

The missing values are the whole numbers between 1 and 16.
Each number is only used once.
Each row is a math equation.
Each column is a math equation.
Remember that multiplication and division are performed before addition and subtraction.

# MATH PUZZLE-09

|   |   |   |   |   |   |   |   |    |
|---|---|---|---|---|---|---|---|----|
| 7 | + |   | + |   | + |   |   | 27 |
| + | ■ | + | ■ | + | ■ | + |   |    |
|   | + |   | + |   | + |   |   | 45 |
| + | ■ | + | ■ | + | ■ | + |   |    |
|   | + |   | + |   | + |   |   | 39 |
| + | ■ | + | ■ | + | ■ | + |   |    |
| 12| + |   | + |   | + |   |   | 25 |
| 50|   | 26|   | 29|   | 31|   |    |

## Fill in the missing numbers

The missing values are the whole numbers between 1 and 16.

Each number is only used once.

Each row is a math equation.

Each column is a math equation.

Remember that multiplication and division are performed before addition and subtraction.

# MATH PUZZLE-10

|   | + |   | + |   | + |   | 38 |
|---|---|---|---|---|---|---|----|
| + |   | + |   | + |   | + |    |
|   | + |   | + |   | + | 3 | 26 |
| + |   | + |   | + |   | + |    |
|   | + |   | + |   | + |   | 26 |
| + |   | + |   | + |   | + |    |
| 7 | + |   | + |   | + |   | 46 |
| 28|   | 50|   | 29|   | 29|    |

**Fill in the missing numbers**

The missing values are the whole numbers between 1 and 16.
Each number is only used once.
Each row is a math equation.
Each column is a math equation.
Remember that multiplication and division are performed before addition and subtraction.

# MATH PUZZLE-11

|   | + | 10 | + |   | + |   | 27 |
|---|---|---|---|---|---|---|---|
| + | ■ | + | ■ | + | ■ | + |   |
|   | + |   | + |   | + |   | 27 |
| + | ■ | + | ■ | + | ■ | + |   |
| 8 | + |   | + |   | + |   | 48 |
| + | ■ | + | ■ | + | ■ | + |   |
|   | + |   | + |   | + |   | 34 |
| 27 |   | 40 |   | 32 |   | 37 |   |

### Fill in the missing numbers

The missing values are the whole numbers between 1 and 16.
Each number is only used once.
Each row is a math equation.
Each column is a math equation.
Remember that multiplication and division are performed before addition and subtraction.

# MATH PUZZLE-12

|   | + |   | + |   | + |   | 30 |
|---|---|---|---|---|---|---|----|
| + | ■ | + | ■ | + | ■ | + |    |
| 8 | + |   | + |   | + |   | 27 |
| + | ■ | + | ■ | + | ■ | + |    |
|   | + |   | + |   | + |   | 36 |
| + | ■ | + | ■ | + | ■ | + |    |
| 10| + |   | + |   | + |   | 43 |
| 26|   | 37|   | 27|   | 46|    |

**Fill in the missing numbers**

The missing values are the whole numbers between 1 and 16.
Each number is only used once.
Each row is a math equation.
Each column is a math equation.
Remember that multiplication and division are performed before addition and subtraction.

# MATH PUZZLE-13

|   | + |   | + |   | + |   | 25 |
| + |   | + |   | + |   | + |    |
| 14 | + |   | + | 15 | + |   | 48 |
| + |   | + |   | + |   | + |    |
|   | + |   | + |   | + |   | 28 |
| + |   | + |   | + |   | + |    |
|   | + |   | + |   | + |   | 35 |
| 38 |   | 30 |   | 45 |   | 23 |    |

**Fill in the missing numbers**

The missing values are the whole numbers between 1 and 16.

Each number is only used once.

Each row is a math equation.

Each column is a math equation.

Remember that multiplication and division are performed before addition and subtraction.

# MATH PUZZLE-14

|   | + |   | + |   | + |   | 38 |
|---|---|---|---|---|---|---|----|
| + | ■ | + | ■ | + | ■ | + |    |
|   | + |   | + |   | + |   | 45 |
| + | ■ | + | ■ | + | ■ | + |    |
|   | + | 1 | + |   | + | 4 | 28 |
| + | ■ | + | ■ | + | ■ | + |    |
|   | + |   | + |   | + |   | 25 |
| 44|   | 29|   | 41|   | 22|    |

**Fill in the missing numbers**

The missing values are the whole numbers between 1 and 16.
Each number is only used once.
Each row is a math equation.
Each column is a math equation.
Remember that multiplication and division are performed before addition and subtraction.

# MATH PUZZLE-15

|   | + |   | + |   | + |   | 41 |
|---|---|---|---|---|---|---|----|
| + |   | + |   | + |   | + |    |
|   | + | 2 | + |   | + |   | 23 |
| + |   | + |   | + |   | + |    |
|   | + |   | + |   | + |   | 34 |
| + |   | + |   | + |   | + |    |
|   | + |   | + |   | + | 6 | 38 |
| 34 |  | 34 |  | 26 |  | 42 |   |

**Fill in the missing numbers**

The missing values are the whole numbers between 1 and 16.

Each number is only used once.

Each row is a math equation.

Each column is a math equation.

Remember that multiplication and division are performed before addition and subtraction.

# MATH PUZZLE-16

|   | −  |   | +  |   | +  |   | 16  |
|---|---|---|---|---|---|---|---|
| − | ■ | − | ■ | − | ■ | + |   |
|   | − |   | − |   | − |   | −10 |
| + | ■ | − | ■ | + | ■ | + |   |
| 4 | − |   | − |   | + |   | 1 |
| + | ■ | − | ■ | − | ■ | + |   |
| 13 | − |   | − |   | + |   | 13 |
| 13 |   | −20 |   | 12 |   | 35 |   |

### Fill in the missing numbers

The missing values are the whole numbers between 1 and 16.

Each number is only used once.

Each row is a math equation.

Each column is a math equation.

Remember that multiplication and division are performed before addition and subtraction.

# MATH PUZZLE-17

## Fill in the missing numbers

The missing values are the whole numbers between 1 and 16.
Each number is only used once.
Each row is a math equation.
Each column is a math equation.
Remember that multiplication and division are performed before addition and subtraction.

# MATH PUZZLE-18

|   | - | 12 | + |   | + |   | 14 |
|---|---|----|---|---|---|---|----|
| - | ■ | -  | ■ | - | ■ | - |    |
|   | + |    | - |   | - | 5 | 7  |
| + | ■ | +  | ■ | - | ■ | + |    |
|   | - |    | + |   | + |   | 11 |
| + | ■ | +  | ■ | - | ■ | - |    |
|   | + |    | - |   | - |   | -2 |
| 18|   | 15 |   |-14|   | -1|    |

## Fill in the missing numbers

The missing values are the whole numbers between 1 and 16.
Each number is only used once.
Each row is a math equation.
Each column is a math equation.
Remember that multiplication and division are performed before addition and subtraction.

# MATH PUZZLE-19

|   | - |   | + |   | - |   | 5 |
|---|---|---|---|---|---|---|---|
| - | ■ | - | ■ | - | ■ | - |   |
|   | - | 2 | - |   | - |   | -22 |
| - | ■ | + | ■ | + | ■ | + |   |
|   | - |   | + |   | + |   | 9 |
| - | ■ | + | ■ | + | ■ | - |   |
|   | + | 13 | + |   | - |   | 16 |
| -8 |   | 20 |   | 4 |   | -8 |   |

**Fill in the missing numbers**

The missing values are the whole numbers between 1 and 16.
Each number is only used once.
Each row is a math equation.
Each column is a math equation.
Remember that multiplication and division are performed before addition and subtraction.

# MATH PUZZLE-20

|   | − |   | − | 13 | − |   | −30 |
|---|---|---|---|----|---|---|-----|
| − |   | − |   | +  |   | − |     |
|   | − |   | − |    | + | 15 | 4  |
| + |   | + |   | −  |   | − |     |
|   | + |   | − |    | + |   | 7   |
| − |   | − |   | −  |   | − |     |
|   | + |   | − |    | + |   | 19  |
| −8 |  | 15 |  | 5 |   | −30 |   |

### Fill in the missing numbers

The missing values are the whole numbers between 1 and 16.
Each number is only used once.
Each row is a math equation.
Each column is a math equation.
Remember that multiplication and division are performed before addition and subtraction.

# MATH PUZZLE-21

| 4 | − | 15 | − |   | − |   | −30 |
|---|---|---|---|---|---|---|---|
| − |   | − |   | − |   | − |   |
|   | + |   | + |   | − |   | 19 |
| − |   | + |   | + |   | − |   |
|   | + |   | − |   | − |   | −7 |
| − |   | + |   | − |   | + |   |
|   | − |   | − |   | + |   | −4 |
| −22 |   | 14 |   | 3 |   | −3 |   |

**Fill in the missing numbers**

The missing values are the whole numbers between 1 and 16.
Each number is only used once.
Each row is a math equation.
Each column is a math equation.
Remember that multiplication and division are performed before addition and subtraction.

# MATH PUZZLE-22

|   | −  | 2 | −  |   | +  |   | 4   |
|---|----|---|----|---|----|---|-----|
| − |    | − |    | + |    | + |     |
|   | +  |   | −  |   | −  |   | 8   |
| − |    | + |    | + |    | + |     |
|   | −  |   | +  |   | −  |   | −19 |
| − |    | + |    | − |    | − |     |
| 9 | −  |   | −  |   | −  |   | −29 |
| −22 |  | 17 |   | 4 |    | 15 |    |

**Fill in the missing numbers**

The missing values are the whole numbers between 1 and 16.
Each number is only used once.
Each row is a math equation.
Each column is a math equation.
Remember that multiplication and division are performed before addition and subtraction.

# MATH PUZZLE-23

|   | - |   | - |   | + |   | -9 |
|---|---|---|---|---|---|---|---|
| - | ■ | - | ■ | - | ■ | + |   |
|   | - |   | + |   | - |   | 10 |
| + | ■ | - | ■ | - | ■ | - |   |
|   | + |   | - | 13 | + |   | 14 |
| - | ■ | - | ■ | - | ■ | + |   |
|   | - | 4 | + |   | - |   | -3 |
| -15 |   | -19 |   | -13 |   | 15 |   |

### Fill in the missing numbers

The missing values are the whole numbers between 1 and 16.

Each number is only used once.

Each row is a math equation.

Each column is a math equation.

Remember that multiplication and division are performed before addition and subtraction.

# MATH PUZZLE-24

## Fill in the missing numbers

The missing values are the whole numbers between 1 and 16.
Each number is only used once.
Each row is a math equation.
Each column is a math equation.
Remember that multiplication and division are performed before addition and subtraction.

# MATH PUZZLE-25

|   |   |   |   |   |   |   |   |
|---|---|---|---|---|---|---|---|
| 6 | − |   | + |   | − |   | 8 |
| − | ■ | − | ■ | + | ■ | + |   |
|   | − |   | − |   | − |   | −3 |
| − | ■ | + | ■ | − | ■ | − |   |
|   | + | 8 | + |   | + |   | 43 |
| + | ■ | + | ■ | − | ■ | − |   |
|   | − |   | + |   | + |   | 16 |
| −12 |   | 7 |   | 7 |   | −14 |   |

**Fill in the missing numbers**

The missing values are the whole numbers between 1 and 16.
Each number is only used once.
Each row is a math equation.
Each column is a math equation.
Remember that multiplication and division are performed before addition and subtraction.

# MATH PUZZLE-26

| 11 | - |   | - |   | + |   | -12 |
|----|---|---|---|---|---|---|-----|
| -  |   | - |   | + |   | - |     |
|    | - |   | + |   | - |   | 17  |
| +  |   | - |   | - |   | + |     |
|    | + |   | + |   | - |   | 15  |
| +  |   | + |   | - |   | + |     |
| 4  | - |   | - |   | - |   | -28 |
| 3  |   | 8 |   | 16|   | 15|     |

**Fill in the missing numbers**

The missing values are the whole numbers between 1 and 16.
Each number is only used once.
Each row is a math equation.
Each column is a math equation.
Remember that multiplication and division are performed before addition and subtraction.

# MATH PUZZLE-27

|   | + |   | + | 6 | + |   | 34 |
|---|---|---|---|---|---|---|---|
| - |   | - |   | - |   | + |   |
|   | + |   | - | 12 | + |   | 21 |
| - |   | - |   | + |   | - |   |
|   | - | 16 | - | 5 | + | 1 | -16 |
| - |   | + |   | + |   | + |   |
|   | + |   | - |   | - |   | -11 |
| -15 |   | -10 |   | 6 |   | 25 |   |

### Fill in the missing numbers

The missing values are the whole numbers between 1 and 16.
Each number is only used once.
Each row is a math equation.
Each column is a math equation.
Remember that multiplication and division are performed before addition and subtraction.

# MATH PUZZLE-28

|   | - |   | - |   | - | 2 | -19 |
|---|---|---|---|---|---|---|-----|
| + | ■ | - | ■ | + | ■ | - |     |
| 8 | + |   | + |   | + |   | 23  |
| - | ■ | - | ■ | + | ■ | - |     |
|   | - | 9 | - |   | + |   | -3  |
| + | ■ | + | ■ | - | ■ | + |     |
|   | + | 15 | - | 1 | - |   | 7   |
| 20 |  | 10 |  | 30 |  | -4 |    |

**Fill in the missing numbers**

The missing values are the whole numbers between 1 and 16.
Each number is only used once.
Each row is a math equation.
Each column is a math equation.
Remember that multiplication and division are performed before addition and subtraction.

# MATH PUZZLE-29

|   | + |   | − |   | + |   | 13 |
| − |   | + |   | − |   | + |   |
|   | − | 14 | − | 4 | + |   | 10 |
| − |   | + |   | + |   | − |   |
|   | − | 7 | + | 15 | + |   | 22 |
| + |   | + |   | + |   | + |   |
|   | − | 10 | + |   | + |   | 3 |
| −15 |   | 42 |   | 26 |   | 23 |   |

**Fill in the missing numbers**

The missing values are the whole numbers between 1 and 16.
Each number is only used once.
Each row is a math equation.
Each column is a math equation.
Remember that multiplication and division are performed before addition and subtraction.

# MATH PUZZLE-30

|   | + |   | − | 9 | + |   | 16 |
|---|---|---|---|---|---|---|----|
| − | ■ | + | ■ | + | ■ | + |    |
| 10 | + | 3 | − |   | + |   | 23 |
| − | ■ | − | ■ | + | ■ | − |    |
|   | − |   | − |   | + | 16 | 12 |
| − | ■ | − | ■ | − | ■ | + |    |
|   | + |   | + | 15 | − |   | 31 |
| −16 |   | −15 |   | −1 |   | 12 |    |

### Fill in the missing numbers

The missing values are the whole numbers between 1 and 16.
Each number is only used once.
Each row is a math equation.
Each column is a math equation.
Remember that multiplication and division are performed before addition and subtraction.

# MATH PUZZLE-31

|   |   |   |   |   |   |   |   |
|---|---|---|---|---|---|---|---|
| 1 | × | 14 | − | 9 | − | 13 | −8 |
| + |   | × |   | + |   | − |   |
|   | − |   | + | 10 | − |   | 10 |
| + |   | − |   | + |   | × |   |
|   | + |   | − |   | + | 7 | 4 |
| × |   | + |   | − |   | − |   |
|   | × |   | − |   | + |   | 49 |
| 29 |   | 93 |   | 15 |   | −17 |   |

### Fill in the missing numbers

The missing values are the whole numbers between 1 and 16.
Each number is only used once.
Each row is a math equation.
Each column is a math equation.
Remember that multiplication and division are performed before addition and subtraction.

# MATH PUZZLE-32

|     |     |     |     |     |     |     |     |
| --- | --- | --- | --- | --- | --- | --- | --- |
| 10  | −   |     | +   |     | −   |     | −4  |
| +   | ■   | ×   | ■   | +   | ■   | ×   |     |
| 6   | −   |     | +   | 7   | +   |     | 8   |
| +   | ■   | −   | ■   | +   | ■   | −   |     |
| 11  | −   |     | +   | 15  | +   | 14  | 39  |
| −   | ■   | −   | ■   | +   | ■   | +   |     |
|     | +   |     | +   |     | −   |     | 21  |
| 14  |     | 122 |     | 38  |     | 1   |     |

## Fill in the missing numbers

The missing values are the whole numbers between 1 and 16.
Each number is only used once.
Each row is a math equation.
Each column is a math equation.
Remember that multiplication and division are performed before addition and subtraction.

# MATH PUZZLE-33

|   | + |   | + |   | − |   | 22 |
|---|---|---|---|---|---|---|---|
| + |   | × |   | + |   | − |   |
| 6 | + | 7 | − |   | × | 2 | 7 |
| − |   | − |   | + |   | − |   |
|   | + |   | + | 14 | − |   | 17 |
| + |   | + |   | − |   | × |   |
| 12 | − | 16 | + |   | − |   | −11 |
| 11 |   | 106 |   | 19 |   | −132 |   |

**Fill in the missing numbers**

The missing values are the whole numbers between 1 and 16.
Each number is only used once.
Each row is a math equation.
Each column is a math equation.
Remember that multiplication and division are performed before addition and subtraction.

# MATH PUZZLE-34

|   | -  |   | +  | 9  | ×  |    | 68 |
|---|----|---|----|----|----|----|----|
| - | ■  | + | ■  | ×  | ■  | ×  |    |
|   | -  | 1 | ×  | 2  | +  |    | 25 |
| - | ■  | - | ■  | +  | ■  | -  |    |
|   | ×  |   | +  | 7  | +  | 13 | 40 |
| × | ■  | - | ■  | +  | ■  | -  |    |
| 3 | ×  |   | +  |    | -  |    | 27 |
| -18 |  | -2 |    | 31 |    | 60 |    |

### Fill in the missing numbers

The missing values are the whole numbers between 1 and 16.
Each number is only used once.
Each row is a math equation.
Each column is a math equation.
Remember that multiplication and division are performed before addition and subtraction.

# MATH PUZZLE-35

|   | + |   | − | 3 | + | 1 | 18 |
|---|---|---|---|---|---|---|---|
| − |   | − |   | − |   | + |   |
|   | + |   | + |   | − |   | 14 |
| − |   | − |   | + |   | + |   |
| 10 | − | 11 | × | 13 | + |   | −117 |
| + |   | + |   | × |   | + |   |
|   | − |   | + | 15 | − |   | 4 |
| 2 |   | −4 |   | 186 |   | 31 |   |

**Fill in the missing numbers**

The missing values are the whole numbers between 1 and 16.
Each number is only used once.
Each row is a math equation.
Each column is a math equation.
Remember that multiplication and division are performed before addition and subtraction.

# MATH PUZZLE-36

|   | + | 7 | − | 12 | + |   | 5 |
|---|---|---|---|---|---|---|---|
| − |   | − |   | + |   | − |   |
|   | + |   | × | 16 | − |   | 138 |
| × |   | − |   | × |   | − |   |
|   | × | 5 | + |   | + |   | 24 |
| + |   | − |   | + |   | − |   |
| 3 | + |   | − | 11 | − |   | −7 |
| 1 |   | −22 |   | 119 |   | −29 |   |

**Fill in the missing numbers**

The missing values are the whole numbers between 1 and 16.
Each number is only used once.
Each row is a math equation.
Each column is a math equation.
Remember that multiplication and division are performed before addition and subtraction.

# MATH PUZZLE-37

|   |   |   |   |   |   |   |   |   |
|---|---|---|---|---|---|---|---|---|
|   | × |   | − |   | + | 16 | 31 |
| × |   | + |   | − |   | + |   |
|   | + |   | × |   | − |   | 59 |
| − |   | × |   | + |   | + |   |
| 5 | + |   | − | 12 | + | 11 | 14 |
| − |   | − |   | × |   | + |   |
| 14 | − |   | + | 3 | + |   | 13 |
| 29 |   | 139 |   | 33 |   | 43 |   |

**Fill in the missing numbers**

The missing values are the whole numbers between 1 and 16.
Each number is only used once.
Each row is a math equation.
Each column is a math equation.
Remember that multiplication and division are performed before addition and subtraction.

# MATH PUZZLE-38

|     | −   |     | +   |     | −   | 6   | 10   |
| --- | --- | --- | --- | --- | --- | --- | ---- |
| ×   |     | −   |     | ×   |     | +   |      |
| 1   | −   | 11  | −   |     | ×   |     | −30  |
| −   |     | −   |     | −   |     | −   |      |
| 8   | +   | 7   | −   |     | ×   |     | −193 |
| −   |     | +   |     | +   |     | ×   |      |
|     | −   | 14  | −   |     | −   |     | −13  |
| −14 |     | −1  |     | 46  |     | −16 |      |

## Fill in the missing numbers

The missing values are the whole numbers between 1 and 16.

Each number is only used once.

Each row is a math equation.

Each column is a math equation.

Remember that multiplication and division are performed before addition and subtraction.

# MATH PUZZLE-39

| 5 | − | 14 | − |  | − |  | −21 |
|---|---|---|---|---|---|---|---|
| + |  | × |  | + |  | − |  |
|  | − |  | − |  | + | 4 | −4 |
| + |  | + |  | × |  | − |  |
|  | + | 2 | + |  | × |  | 132 |
| + |  | − |  | − |  | × |  |
|  | × | 16 | − |  | + | 13 | 202 |
| 34 |  | 112 |  | 56 |  | −198 |  |

### Fill in the missing numbers

The missing values are the whole numbers between 1 and 16.

Each number is only used once.

Each row is a math equation.

Each column is a math equation.

Remember that multiplication and division are performed before addition and subtraction.

# MATH PUZZLE-40

|   |   |   |   |   |   |   |   |
|---|---|---|---|---|---|---|---|
| 12 | − |    | − | 9  | + |    | 4 |
| ×  | ■ | +  | ■ | −  | ■ | −  |   |
|    | − | 16 | × |    | + |    | 2 |
| −  | ■ | +  | ■ | +  | ■ | +  |   |
| 4  | + | 13 | − |    | − | 3  | 7 |
| +  | ■ | −  | ■ | −  | ■ | −  |   |
|    | + |    | + |    | − |    | 20 |
| 118 |   | 20 |   | 0  |   | −10 |   |

**Fill in the missing numbers**

The missing values are the whole numbers between 1 and 16.
Each number is only used once.
Each row is a math equation.
Each column is a math equation.
Remember that multiplication and division are performed before addition and subtraction.

# MATH PUZZLE-41

|   | + | 14 | + |   | − |   | 28 |
|---|---|----|---|---|---|---|----|
| + |   | +  |   | × |   | − |    |
|   | × | 11 | − | 7 | − |   | 25 |
| × |   | −  |   | + |   | ÷ |    |
|   | − |    | + | 13 | − | 3 | 9 |
| + |   | ×  |   | + |   | × |    |
|   | × | 10 | − |   | − | 16 | 66 |
| 44 |  | −35 |  | 28 |  | −62 |   |

### Fill in the missing numbers

The missing values are the whole numbers between 1 and 16.

Each number is only used once.

Each row is a math equation.

Each column is a math equation.

Remember that multiplication and division are performed before addition and subtraction.

# MATH PUZZLE-42

|   | ×  | 8  | +  | 1  | +  | 11 | 116 |
|---|----|----|----|----|----|----|-----|
| + |    | −  |    | +  |    | −  |     |
| 12| +  |    | +  |    | ×  | 6  | 77  |
| − |    | +  |    | +  |    | ÷  |     |
|   | ×  | 15 | +  |    | ÷  | 2  | 113 |
| + |    | ÷  |    | −  |    | −  |     |
|   | −  |    | ×  |    | −  |    | −17 |
| 32|    | 8  |    | 18 |    | 4  |     |

**Fill in the missing numbers**

The missing values are the whole numbers between 1 and 16.

Each number is only used once.

Each row is a math equation.

Each column is a math equation.

Remember that multiplication and division are performed before addition and subtraction.

# MATH PUZZLE-43

|   | - | 8 | - |   | + | 9 | 11 |
|---|---|---|---|---|---|---|---|
| - | ■ | - | ■ | - | ■ | + |   |
| 14 | × | 5 | + |   | + |   | 96 |
| ÷ | ■ | - | ■ | - | ■ | - |   |
|   | - | 4 | × |   | - | 1 | -59 |
| - | ■ | + | ■ | + | ■ | × |   |
|   | + | 12 | ÷ |   | - |   | 6 |
| -5 |   | 11 |   | -16 |   | 18 |   |

### Fill in the missing numbers

The missing values are the whole numbers between 1 and 16.
Each number is only used once.
Each row is a math equation.
Each column is a math equation.
Remember that multiplication and division are performed before addition and subtraction.

# MATH PUZZLE-44

|   | + |   | + |   | − |   | 30 |
|---|---|---|---|---|---|---|---|
| − |   | − |   | − |   | + |   |
| 1 | + |   | + |   | + |   | 38 |
| + |   | × |   | × |   | × |   |
| 2 | + | 12 | × |   | − | 14 | 48 |
| − |   | ÷ |   | − |   | ÷ |   |
| 3 | − | 4 | − | 9 | + |   | −3 |
| 8 |   | −9 |   | −63 |   | 38 |   |

### Fill in the missing numbers

The missing values are the whole numbers between 1 and 16.
Each number is only used once.
Each row is a math equation.
Each column is a math equation.
Remember that multiplication and division are performed before addition and subtraction.

# MATH PUZZLE-45

|   | + |   | + | 13 | + | 8 | 29 |
|---|---|---|---|----|---|---|----|
| - |   | - |   | -  |   | + |    |
| 7 | - | 16| + | 1  | - |   | -22|
| + |   | - |   | -  |   | - |    |
| 2 | × |   | - |    | - |   | -3 |
| + |   | × |   | ×  |   | + |    |
|   | - |   | + | 12 | + |   | 29 |
| 10|   |-49|   |-60 |   | 18|    |

### Fill in the missing numbers

The missing values are the whole numbers between 1 and 16.
Each number is only used once.
Each row is a math equation.
Each column is a math equation.
Remember that multiplication and division are performed before addition and subtraction.

# MATH PUZZLE-46

|   | − |   | ÷ |   | − |   | −17 |
|---|---|---|---|---|---|---|---|
| + | ■ | × | ■ | + | ■ | − |   |
|   | + | 15 | + | 7 | + |   | 34 |
| × | ■ | ÷ | ■ | + | ■ | − |   |
| 14 | + | 5 | + | 10 | − | 6 | 23 |
| − | ■ | − | ■ | + | ■ | − |   |
| 8 | − |   | + |   | + |   | 8 |
| 122 |   | 36 |   | 20 |   | −7 |   |

## Fill in the missing numbers

The missing values are the whole numbers between 1 and 16.
Each number is only used once.
Each row is a math equation.
Each column is a math equation.
Remember that multiplication and division are performed before addition and subtraction.

# MATH PUZZLE-47

|   | + | 3 | − |   | − | 9 | −19 |
|---|---|---|---|---|---|---|---|
| − |   | + |   | − |   | + |   |
|   | − | 7 | + | 14 | + |   | 25 |
| − |   | − |   | + |   | + |   |
|   | + | 12 | + | 1 | + |   | 30 |
| + |   | × |   | + |   | + |   |
|   | + |   | × |   | − | 16 | 55 |
| −7 |   | −122 |   | 8 |   | 46 |   |

**Fill in the missing numbers**

The missing values are the whole numbers between 1 and 16.
Each number is only used once.
Each row is a math equation.
Each column is a math equation.
Remember that multiplication and division are performed before addition and subtraction.

# MATH PUZZLE-48

|   | - |   | + | 16 | - | 12 | 7 |
|---|---|---|---|----|---|----|---|
| + | ■ | × | ■ | × | ■ | ÷ |   |
| 7 | - |   | + | 5 | + |   | 9 |
| - | ■ | + | ■ | + | ■ | ÷ |   |
| 8 | - |   | - |   | ÷ | 3 | -11 |
| ÷ | ■ | + | ■ | + | ■ | - |   |
|   | - | 13 | - |   | + |   | -10 |
| 7 |   | 36 |   | 105 |   | -11 |   |

**Fill in the missing numbers**

The missing values are the whole numbers between 1 and 16.
Each number is only used once.
Each row is a math equation.
Each column is a math equation.
Remember that multiplication and division are performed before addition and subtraction.

# MATH PUZZLE-49

|  |  |  |  |  |  |  |  |
|---|---|---|---|---|---|---|---|
| 15 | × | 7 | − |  | − | 9 | 92 |
| + |  | + |  | × |  | − |  |
| 10 | − |  | ÷ | 8 | × |  | −16 |
| + |  | − |  | + |  | × |  |
|  | + |  | − |  | + |  | 13 |
| × |  | − |  | − |  | + |  |
|  | + |  | + | 14 | + | 1 | 29 |
| 47 |  | 8 |  | 24 |  | −55 |  |

**Fill in the missing numbers**

The missing values are the whole numbers between 1 and 16.
Each number is only used once.
Each row is a math equation.
Each column is a math equation.
Remember that multiplication and division are performed before addition and subtraction.

# MATH PUZZLE-50

|  |  |  |  |  |  |  |  |
|---|---|---|---|---|---|---|---|
| 3 | − |  | − | 11 | × | 2 | −20 |
| − |  | + |  | × |  | − |  |
|  | − | 15 | × | 13 | − |  | −191 |
| × |  | − |  | + |  | − |  |
|  | + | 7 | + | 6 | − |  | 14 |
| + |  | × |  | + |  | + |  |
|  | + |  | − |  | + |  | 13 |
| −119 |  | −96 |  | 161 |  | −11 |  |

### Fill in the missing numbers

The missing values are the whole numbers between 1 and 16.

Each number is only used once.

Each row is a math equation.

Each column is a math equation.

Remember that multiplication and division are performed before addition and subtraction.

# Solution

## MATH PUZZLE-01

| 2 | + | 12 | + | 3 | + | 9 | 26 |
|---|---|---|---|---|---|---|---|
| + | | + | | + | | + | |
| 1 | + | 7 | + | 16 | + | 11 | 35 |
| + | | + | | + | | + | |
| 10 | + | 4 | + | 6 | + | 8 | 28 |
| + | | + | | + | | + | |
| 15 | + | 14 | + | 13 | + | 5 | 47 |
| 28 | | 37 | | 38 | | 33 | |

## MATH PUZZLE-02

| 2 | + | 9 | + | 10 | + | 7 | 28 |
|---|---|---|---|---|---|---|---|
| + | | + | | + | | + | |
| 8 | + | 11 | + | 1 | + | 15 | 35 |
| + | | + | | + | | + | |
| 12 | + | 13 | + | 16 | + | 14 | 55 |
| + | | + | | + | | + | |
| 6 | + | 4 | + | 5 | + | 3 | 18 |
| 28 | | 37 | | 32 | | 39 | |

## MATH PUZZLE-03

| 1 | + | 3 | + | 12 | + | 5 | 21 |
|---|---|---|---|----|---|---|----|
| + |   | + |   | +  |   | + |    |
| 10 | + | 6 | + | 7 | + | 4 | 27 |
| + |   | + |   | + |   | + |    |
| 2 | + | 8 | + | 15 | + | 16 | 41 |
| + |   | + |   | +  |   | +  |    |
| 9 | + | 13 | + | 14 | + | 11 | 47 |
| 22 |  | 30 |  | 48 |  | 36 |  |

## MATH PUZZLE-04

| 8 | + | 9 | + | 7 | + | 4 | 28 |
|---|---|---|---|---|---|---|----|
| + |   | + |   | + |   | + |    |
| 5 | + | 12 | + | 14 | + | 15 | 46 |
| + |   | +  |   | +  |   | +  |    |
| 3 | + | 10 | + | 2 | + | 1 | 16 |
| + |   | +  |   | + |   | + |    |
| 16 | + | 11 | + | 6 | + | 13 | 46 |
| 32 |  | 42 |  | 29 |  | 33 |   |

## MATH PUZZLE-05

| 5 | + | 15 | + | 9 | + | 2 | 31 |
|---|---|---|---|---|---|---|---|
| + |   | + |   | + |   | + |   |
| 1 | + | 6 | + | 8 | + | 7 | 22 |
| + |   | + |   | + |   | + |   |
| 13 | + | 14 | + | 10 | + | 16 | 53 |
| + |   | + |   | + |   | + |   |
| 4 | + | 11 | + | 12 | + | 3 | 30 |
| 23 |   | 46 |   | 39 |   | 28 |   |

## MATH PUZZLE-06

| 11 | + | 7 | + | 14 | + | 1 | 33 |
|---|---|---|---|---|---|---|---|
| + |   | + |   | + |   | + |   |
| 16 | + | 15 | + | 2 | + | 5 | 38 |
| + |   | + |   | + |   | + |   |
| 4 | + | 10 | + | 6 | + | 8 | 28 |
| + |   | + |   | + |   | + |   |
| 12 | + | 3 | + | 13 | + | 9 | 37 |
| 43 |   | 35 |   | 35 |   | 23 |   |

## MATH PUZZLE-07

| 4 | + | 6 | + | 12 | + | 9 | 31 |
|---|---|---|---|---|---|---|---|
| + |   | + |   | + |   | + |   |
| 3 | + | 2 | + | 14 | + | 8 | 27 |
| + |   | + |   | + |   | + |   |
| 1 | + | 11 | + | 16 | + | 13 | 41 |
| + |   | + |   | + |   | + |   |
| 15 | + | 10 | + | 7 | + | 5 | 37 |
| 23 |   | 29 |   | 49 |   | 35 |   |

## MATH PUZZLE-08

| 1 | + | 13 | + | 11 | + | 6 | 31 |
|---|---|---|---|---|---|---|---|
| + |   | + |   | + |   | + |   |
| 12 | + | 15 | + | 3 | + | 14 | 44 |
| + |   | + |   | + |   | + |   |
| 10 | + | 8 | + | 7 | + | 5 | 30 |
| + |   | + |   | + |   | + |   |
| 16 | + | 9 | + | 4 | + | 2 | 31 |
| 39 |   | 45 |   | 25 |   | 27 |   |

## MATH PUZZLE-09

| 7 | + | 5 | + | 9 | + | 6 | 27 |
|---|---|---|---|---|---|---|---|
| + |   | + |   | + |   | + |   |
| 16 | + | 14 | + | 2 | + | 13 | 45 |
| + |   | + |   | + |   | + |   |
| 15 | + | 3 | + | 10 | + | 11 | 39 |
| + |   | + |   | + |   | + |   |
| 12 | + | 4 | + | 8 | + | 1 | 25 |
| 50 |   | 26 |   | 29 |   | 31 |   |

## MATH PUZZLE-10

| 15 | + | 10 | + | 8 | + | 5 | 38 |
|---|---|---|---|---|---|---|---|
| + |   | + |   | + |   | + |   |
| 4 | + | 13 | + | 6 | + | 3 | 26 |
| + |   | + |   | + |   | + |   |
| 2 | + | 11 | + | 1 | + | 12 | 26 |
| + |   | + |   | + |   | + |   |
| 7 | + | 16 | + | 14 | + | 9 | 46 |
| 28 |   | 50 |   | 29 |   | 29 |   |

## MATH PUZZLE-11

| 11 | + | 10 | + | 2 | + | 4 | 27 |
|---|---|---|---|---|---|---|---|
| + |   | + |   | + |   | + |   |
| 3 | + | 9 | + | 1 | + | 14 | 27 |
| + |   | + |   | + |   | + |   |
| 8 | + | 15 | + | 13 | + | 12 | 48 |
| + |   | + |   | + |   | + |   |
| 5 | + | 6 | + | 16 | + | 7 | 34 |
| 27 |   | 40 |   | 32 |   | 37 |   |

## MATH PUZZLE-12

| 1 | + | 5 | + | 9 | + | 15 | 30 |
|---|---|---|---|---|---|---|---|
| + |   | + |   | + |   | + |   |
| 8 | + | 6 | + | 11 | + | 2 | 27 |
| + |   | + |   | + |   | + |   |
| 7 | + | 12 | + | 4 | + | 13 | 36 |
| + |   | + |   | + |   | + |   |
| 10 | + | 14 | + | 3 | + | 16 | 43 |
| 26 |   | 37 |   | 27 |   | 46 |   |

## MATH PUZZLE-13

| 1 | + | 9 | + | 13 | + | 2 | 25 |
|---|---|---|---|---|---|---|---|
| + |   | + |   | + |   | + |   |
| 14 | + | 11 | + | 15 | + | 8 | 48 |
| + |   | + |   | + |   | + |   |
| 7 | + | 6 | + | 12 | + | 3 | 28 |
| + |   | + |   | + |   | + |   |
| 16 | + | 4 | + | 5 | + | 10 | 35 |
| 38 |   | 30 |   | 45 |   | 23 |   |

## MATH PUZZLE-14

| 12 | + | 9 | + | 7 | + | 10 | 38 |
|---|---|---|---|---|---|---|---|
| + |   | + |   | + |   | + |   |
| 13 | + | 14 | + | 16 | + | 2 | 45 |
| + |   | + |   | + |   | + |   |
| 8 | + | 1 | + | 15 | + | 4 | 28 |
| + |   | + |   | + |   | + |   |
| 11 | + | 5 | + | 3 | + | 6 | 25 |
| 44 |   | 29 |   | 41 |   | 22 |   |

## MATH PUZZLE-15

| 14 | + | 15 | + | 1 | + | 11 | 41 |
|---|---|---|---|---|---|---|---|
| + |   | + |   | + |   | + |   |
| 4 | + | 2 | + | 8 | + | 9 | 23 |
| + |   | + |   | + |   | + |   |
| 3 | + | 5 | + | 10 | + | 16 | 34 |
| + |   | + |   | + |   | + |   |
| 13 | + | 12 | + | 7 | + | 6 | 38 |
| 34 |   | 34 |   | 26 |   | 42 |   |

## MATH PUZZLE-16

| 10 | - | 5 | + | 9 | + | 2 | 16 |
|---|---|---|---|---|---|---|---|
| - |   | - |   | - |   | + |   |
| 14 | - | 16 | - | 1 | - | 7 | -10 |
| + |   | - |   | + |   | + |   |
| 4 | - | 6 | - | 12 | + | 15 | 1 |
| + |   | - |   | - |   | + |   |
| 13 | - | 3 | - | 8 | + | 11 | 13 |
| 13 |   | -20 |   | 12 |   | 35 |   |

## MATH PUZZLE-17

| 5 | − | 14 | + | 2 | + | 13 | 6 |
|---|---|---|---|---|---|---|---|
| − |   | − |   | + |   | − |   |
| 9 | + | 16 | + | 4 | − | 6 | 23 |
| + |   | − |   | + |   | + |   |
| 12 | + | 3 | + | 15 | + | 7 | 37 |
| − |   | − |   | − |   | − |   |
| 8 | − | 10 | + | 11 | + | 1 | 10 |
| 0 |   | −15 |   | 10 |   | 13 |   |

## MATH PUZZLE-18

| 4 | − | 12 | + | 14 | + | 8 | 14 |
|---|---|---|---|---|---|---|---|
| − |   | − |   | − |   | − |   |
| 2 | + | 13 | − | 3 | − | 5 | 7 |
| + |   | + |   | − |   | + |   |
| 1 | − | 6 | + | 9 | + | 7 | 11 |
| + |   | + |   | − |   | − |   |
| 15 | + | 10 | − | 16 | − | 11 | −2 |
| 18 |   | 15 |   | −14 |   | −1 |   |

## MATH PUZZLE-19

| 15 | - | 4 | + | 8 | - | 14 | 5 |
|---|---|---|---|---|---|---|---|
| - |  | - |  | - |  | - | |
| 7 | - | 2 | - | 16 | - | 11 | -22 |
| - |  | + |  | + |  | + | |
| 10 | - | 5 | + | 3 | + | 1 | 9 |
| - |  | + |  | + |  | - | |
| 6 | + | 13 | + | 9 | - | 12 | 16 |
| -8 |  | 20 |  | 4 |  | -8 | |

## MATH PUZZLE-20

| 1 | - | 14 | - | 13 | - | 4 | -30 |
|---|---|---|---|---|---|---|---|
| - |  | - |  | + |  | - | |
| 6 | - | 5 | - | 12 | + | 15 | 4 |
| + |  | + |  | - |  | - | |
| 7 | + | 8 | - | 11 | + | 3 | 7 |
| - |  | - |  | - |  | - | |
| 10 | + | 2 | - | 9 | + | 16 | 19 |
| -8 |  | 15 |  | 5 |  | -30 | |

## MATH PUZZLE-21

| 4 | − | 15 | − | 16 | − | 3 | −30 |
|---|---|---|---|---|---|---|---|
| − |  | − |  | − |  | − |  |
| 7 | + | 10 | + | 11 | − | 9 | 19 |
| − |  | + |  | + |  | − |  |
| 6 | + | 1 | − | 12 | − | 2 | −7 |
| − |  | + |  | − |  | + |  |
| 13 | − | 8 | − | 14 | + | 5 | −4 |
| −22 |  | 14 |  | 3 |  | −3 |  |

## MATH PUZZLE-22

| 5 | − | 2 | − | 6 | + | 7 | 4 |
|---|---|---|---|---|---|---|---|
| − |  | − |  | + |  | + |  |
| 15 | + | 10 | − | 13 | − | 4 | 8 |
| − |  | + |  | + |  | + |  |
| 3 | − | 11 | + | 1 | − | 12 | −19 |
| − |  | + |  | − |  | − |  |
| 9 | − | 14 | − | 16 | − | 8 | −29 |
| −22 |  | 17 |  | 4 |  | 15 |  |

## MATH PUZZLE-23

| 3 | − | 6 | − | 15 | + | 9 | -9 |
|---|---|---|---|---|---|---|---|
| − |   | − |   | − |   | + |   |
| 8 | − | 7 | + | 10 | − | 1 | 10 |
| + |   | − |   | − |   | − |   |
| 2 | + | 14 | − | 13 | + | 11 | 14 |
| − |   | − |   | − |   | + |   |
| 12 | − | 4 | + | 5 | − | 16 | -3 |
| -15 |   | -19 |   | -13 |   | 15 |   |

## MATH PUZZLE-24

| 5 | + | 6 | − | 10 | − | 14 | -13 |
|---|---|---|---|---|---|---|---|
| − |   | − |   | + |   | + |   |
| 13 | + | 8 | − | 4 | − | 3 | 14 |
| − |   | − |   | + |   | + |   |
| 16 | − | 15 | − | 9 | + | 7 | -1 |
| − |   | + |   | + |   | − |   |
| 11 | + | 2 | + | 12 | + | 1 | 26 |
| -35 |   | -15 |   | 35 |   | 23 |   |

## MATH PUZZLE-25

| 6 | − | 1 | + | 16 | − | 13 | 8 |
|---|---|---|---|---|---|---|---|
| − |  | − |  | + |  | + |  |
| 12 | − | 9 | − | 4 | − | 2 | −3 |
| − |  | + |  | − |  | − |  |
| 11 | + | 8 | + | 10 | + | 14 | 43 |
| + |  | + |  | − |  | − |  |
| 5 | − | 7 | + | 3 | + | 15 | 16 |
| −12 |  | 7 |  | 7 |  | −14 |  |

## MATH PUZZLE-26

| 11 | − | 12 | − | 16 | + | 5 | −12 |
|---|---|---|---|---|---|---|---|
| − |  | − |  | + |  | − |  |
| 13 | − | 3 | + | 14 | − | 7 | 17 |
| + |  | − |  | − |  | + |  |
| 1 | + | 10 | + | 6 | − | 2 | 15 |
| + |  | + |  | − |  | + |  |
| 4 | − | 9 | − | 8 | − | 15 | −28 |
| 3 |  | 8 |  | 16 |  | 15 |  |

# MATH PUZZLE-27

| 10 | + | 15 | + | 6 | + | 3 | 34 |
|---|---|---|---|---|---|---|---|
| − | ■ | − | ■ | − | ■ | + | |
| 13 | + | 11 | − | 12 | + | 9 | 21 |
| − | ■ | − | ■ | + | ■ | − | |
| 4 | − | 16 | − | 5 | + | 1 | −16 |
| − | ■ | + | ■ | + | ■ | + | |
| 8 | + | 2 | − | 7 | − | 14 | −11 |
| −15 | | −10 | | 6 | | 25 | |

# MATH PUZZLE-28

| 10 | − | 11 | − | 16 | − | 2 | −19 |
|---|---|---|---|---|---|---|---|
| + | ■ | − | ■ | + | ■ | − | |
| 8 | + | 7 | + | 3 | + | 5 | 23 |
| − | ■ | − | ■ | + | ■ | − | |
| 4 | − | 9 | − | 12 | + | 14 | −3 |
| + | ■ | + | ■ | − | ■ | + | |
| 6 | + | 15 | − | 1 | − | 13 | 7 |
| 20 | | 10 | | 30 | | −4 | |

## MATH PUZZLE-29

| 8 | + | 11 | − | 9 | + | 3 | 13 |
|---|---|---|---|---|---|---|---|
| − | ■ | + | ■ | − | ■ | + | |
| 12 | − | 14 | − | 4 | + | 16 | 10 |
| − | ■ | + | ■ | + | ■ | − | |
| 13 | − | 7 | + | 15 | + | 1 | 22 |
| + | ■ | + | ■ | + | ■ | + | |
| 2 | − | 10 | + | 6 | + | 5 | 3 |
| −15 | | 42 | | 26 | | 23 | |

## MATH PUZZLE-30

| 7 | + | 6 | − | 9 | + | 12 | 16 |
|---|---|---|---|---|---|---|---|
| − | ■ | + | ■ | + | ■ | + | |
| 10 | + | 3 | − | 4 | + | 14 | 23 |
| − | ■ | − | ■ | + | ■ | − | |
| 8 | − | 11 | − | 1 | + | 16 | 12 |
| − | ■ | − | ■ | − | ■ | + | |
| 5 | + | 13 | + | 15 | − | 2 | 31 |
| −16 | | −15 | | −1 | | 12 | |

## MATH PUZZLE-31

| 1 | × | 14 | - | 9 | - | 13 | -8 |
|---|---|----|---|---|---|----|----|
| + |   | ×  |   | + |   | -  |    |
| 8 | - | 6  | + | 10| - | 2  | 10 |
| + |   | -  |   | + |   | ×  |    |
| 5 | + | 3  | - | 11| + | 7  | 4  |
| × |   | +  |   | - |   | -  |    |
| 4 | × | 12 | - | 15| + | 16 | 49 |
| 29|   | 93 |   | 15|   | -17|    |

## MATH PUZZLE-32

| 10 | - | 16 | + | 4 | - | 2  | -4 |
|----|---|----|---|---|---|----|----|
| +  |   | ×  |   | + |   | ×  |    |
| 6  | - | 8  | + | 7 | + | 3  | 8  |
| +  |   | -  |   | + |   | -  |    |
| 11 | - | 1  | + | 15| + | 14 | 39 |
| -  |   | -  |   | + |   | +  |    |
| 13 | + | 5  | + | 12| - | 9  | 21 |
| 14 |   | 122|   | 38|   | 1  |    |

## MATH PUZZLE-33

| | | | | | | | |
|---|---|---|---|---|---|---|---|
| 4 | + | 13 | + | 10 | − | 5 | 22 |
| + | | × | | + | | − | |
| 6 | + | 7 | − | 3 | × | 2 | 7 |
| − | | − | | + | | − | |
| 11 | + | 1 | + | 14 | − | 9 | 17 |
| + | | + | | | | × | |
| 12 | − | 16 | + | 8 | − | 15 | −11 |
| 11 | | 106 | | 19 | | −132 | |

## MATH PUZZLE-34

| | | | | | | | |
|---|---|---|---|---|---|---|---|
| 10 | − | 14 | + | 9 | × | 8 | 68 |
| − | | + | | × | | × | |
| 16 | − | 1 | × | 2 | + | 11 | 25 |
| − | | − | | + | | − | |
| 4 | × | 5 | + | 7 | + | 13 | 40 |
| × | | − | | + | | − | |
| 3 | × | 12 | + | 6 | − | 15 | 27 |
| −18 | | −2 | | 31 | | 60 | |

## MATH PUZZLE-35

| 14 | + | 6 | − | 3 | + | 1 | 18 |
|---|---|---|---|---|---|---|---|
| − |   | − |   | − |   | + |   |
| 4 | + | 7 | + | 12 | − | 9 | 14 |
| − |   | − |   | + |   | + |   |
| 10 | − | 11 | × | 13 | + | 16 | −117 |
| + |   | + |   | × |   | + |   |
| 2 | − | 8 | + | 15 | − | 5 | 4 |
| 2 |   | −4 |   | 186 |   | 31 |   |

## MATH PUZZLE-36

| 2 | + | 7 | − | 12 | + | 8 | 5 |
|---|---|---|---|---|---|---|---|
| − |   | − |   | + |   | − |   |
| 4 | + | 9 | × | 16 | − | 10 | 138 |
| × |   | − |   | × |   | − |   |
| 1 | × | 5 | + | 6 | + | 13 | 24 |
| + |   | − |   | + |   | − |   |
| 3 | + | 15 | − | 11 | − | 14 | −7 |
| 1 |   | −22 |   | 119 |   | −29 |   |

## MATH PUZZLE-37

| 8 | × | 2 | − | 1 | + | 16 | 31 |
|---|---|---|---|---|---|---|---|
| × |   | + |   | − |   | + |   |
| 6 | + | 15 | × | 4 | − | 7 | 59 |
| − |   | × |   | + |   | + |   |
| 5 | + | 10 | − | 12 | + | 11 | 14 |
| − |   | − |   | × |   | + |   |
| 14 | − | 13 | + | 3 | + | 9 | 13 |
| 29 |   | 139 |   | 33 |   | 43 |   |

## MATH PUZZLE-38

| 9 | − | 3 | + | 10 | − | 6 | 10 |
|---|---|---|---|---|---|---|---|
| × |   | − |   | × |   | + |   |
| 1 | − | 11 | − | 5 | × | 4 | −30 |
| − |   | − |   | − |   | − |   |
| 8 | + | 7 | − | 16 | × | 13 | −193 |
| − |   | + |   | + |   | × |   |
| 15 | − | 14 | − | 12 | − | 2 | −13 |
| −14 |   | −1 |   | 46 |   | −16 |   |

## MATH PUZZLE-39

| 5 | − | 14 | − | 11 | − | 1 | -21 |
|---|---|---|---|---|---|---|---|
| + |  | × |  | + |  | − |  |
| 7 | − | 9 | − | 6 | + | 4 | -4 |
| + |  | + |  | × |  | − |  |
| 10 | + | 2 | + | 8 | × | 15 | 132 |
| + |  | − |  | − |  | × |  |
| 12 | × | 16 | − | 3 | + | 13 | 202 |
| 34 |  | 112 |  | 56 |  | -198 |  |

## MATH PUZZLE-40

| 12 | − | 5 | − | 9 | + | 6 | 4 |
|---|---|---|---|---|---|---|---|
| × |  | + |  | − |  | − |  |
| 10 | − | 16 | × | 1 | + | 8 | 2 |
| − |  | + |  | + |  | + |  |
| 4 | + | 13 | − | 7 | − | 3 | 7 |
| + |  | − |  | − |  | − |  |
| 2 | + | 14 | + | 15 | − | 11 | 20 |
| 118 |  | 20 |  | 0 |  | -10 |  |

## MATH PUZZLE-41

| 15 | + | 14 | + | 1 | − | 2 | 28 |
|---|---|---|---|---|---|---|---|
| + | ■ | + | ■ | × | ■ | − | |
| 4 | × | 11 | − | 7 | − | 12 | 25 |
| × | ■ | − | ■ | + | ■ | ÷ | |
| 5 | − | 6 | + | 13 | − | 3 | 9 |
| + | ■ | × | ■ | + | ■ | × | |
| 9 | × | 10 | − | 8 | − | 16 | 66 |
| 44 | | −35 | | 28 | | −62 | |

## MATH PUZZLE-42

| 13 | × | 8 | + | 1 | + | 11 | 116 |
|---|---|---|---|---|---|---|---|
| + | ■ | − | ■ | + | ■ | − | |
| 12 | + | 5 | + | 10 | × | 6 | 77 |
| − | ■ | + | ■ | + | ■ | ÷ | |
| 7 | × | 15 | + | 16 | ÷ | 2 | 113 |
| + | ■ | ÷ | ■ | − | ■ | − | |
| 14 | − | 3 | × | 9 | − | 4 | −17 |
| 32 | | 8 | | 18 | | 4 | |

## MATH PUZZLE-43

| 13 | − | 8 | − | 3 | + | 9 | 11 |
|---|---|---|---|---|---|---|---|
| − |  | − |  | − |  | + |  |
| 14 | × | 5 | + | 10 | + | 16 | 96 |
| ÷ |  | − |  |  |  | − |  |
| 2 | − | 4 | × | 15 | − | 1 | −59 |
| − |  | + |  | + |  | × |  |
| 11 | + | 12 | ÷ | 6 | − | 7 | 6 |
| −5 |  | 11 |  | −16 |  | 18 |  |

## MATH PUZZLE-44

| 10 | + | 15 | + | 11 | − | 6 | 30 |
|---|---|---|---|---|---|---|---|
| − |  | − |  | − |  | + |  |
| 1 | + | 8 | + | 13 | + | 16 | 38 |
| + |  | × |  | × |  | × |  |
| 2 | + | 12 | × | 5 | − | 14 | 48 |
| − |  | ÷ |  | − |  | ÷ |  |
| 3 | − | 4 | − | 9 | + | 7 | −3 |
| 8 |  | −9 |  | −63 |  | 38 |  |

## MATH PUZZLE-45

| 5 | + | 3 | + | 13 | + | 8 | 29 |
|---|---|---|---|---|---|---|---|
| - |   | - |   | - |   | + |   |
| 7 | - | 16 | + | 1 | - | 14 | -22 |
| + |   | - |   | - |   | - |   |
| 2 | × | 9 | - | 6 | - | 15 | -3 |
| + |   | × |   | × |   | + |   |
| 10 | - | 4 | + | 12 | + | 11 | 29 |
| 10 |   | -49 |   | -60 |   | 18 |   |

## MATH PUZZLE-46

| 4 | - | 16 | ÷ | 2 | - | 13 | -17 |
|---|---|---|---|---|---|---|---|
| + |   | × |   | + |   | - |   |
| 9 | + | 15 | + | 7 | + | 3 | 34 |
| × |   | ÷ |   | + |   | - |   |
| 14 | + | 5 | + | 10 | - | 6 | 23 |
| - |   | - |   | + |   | - |   |
| 8 | - | 12 | + | 1 | + | 11 | 8 |
| 122 |   | 36 |   | 20 |   | -7 |   |

## MATH PUZZLE-47

| | | | | | | | |
|---|---|---|---|---|---|---|---|
| 2 | + | 3 | - | 15 | - | 9 | -19 |
| - | | + | | - | | + | |
| 10 | - | 7 | + | 14 | + | 8 | 25 |
| - | | - | | + | | + | |
| 4 | + | 12 | + | 1 | + | 13 | 30 |
| + | | × | | + | | + | |
| 5 | + | 11 | × | 6 | - | 16 | 55 |
| -7 | | -122 | | 8 | | 46 | |

## MATH PUZZLE-48

| | | | | | | | |
|---|---|---|---|---|---|---|---|
| 4 | - | 1 | + | 16 | - | 12 | 7 |
| + | | × | | × | | ÷ | |
| 7 | - | 9 | + | 5 | + | 6 | 9 |
| - | | + | | + | | ÷ | |
| 8 | - | 14 | - | 15 | ÷ | 3 | -11 |
| ÷ | | + | | + | | - | |
| 2 | - | 13 | - | 10 | + | 11 | -10 |
| 7 | | 36 | | 105 | | -11 | |

## MATH PUZZLE-49

| 15 | × | 7 | - | 4 | - | 9 | 92 |
|---|---|---|---|---|---|---|---|
| + |  | + |  | × |  | - |  |
| 10 | - | 16 | ÷ | 8 | × | 13 | -16 |
| + |  | - |  | + |  | × |  |
| 11 | + | 3 | - | 6 | + | 5 | 13 |
| × |  | - |  | - |  | + |  |
| 2 | + | 12 | + | 14 | + | 1 | 29 |
| 47 |  | 8 |  | 24 |  | -55 |  |

## MATH PUZZLE-50

| 3 | - | 1 | - | 11 | × | 2 | -20 |
|---|---|---|---|---|---|---|---|
| - |  | + |  | × |  | - |  |
| 14 | - | 15 | × | 13 | - | 10 | -191 |
| × |  | - |  | + |  | - |  |
| 9 | + | 7 | + | 6 | - | 8 | 14 |
| + |  | × |  | + |  | + |  |
| 4 | + | 16 | - | 12 | + | 5 | 13 |
| -119 |  | -96 |  | 161 |  | -11 |  |

www.ingramcontent.com/pod-product-compliance
Lightning Source LLC
Chambersburg PA
CBHW070209230526
45471CB00002B/891